Nasir WOUYIE

PORT OVERCAPACITY AND UNDERACTIVITY

Nasir WOUYIE

PORT OVERCAPACITY AND UNDERACTIVITY

ScienciaScripts

Cover image: www.ingimage.com

This book is a translation from the original published under ISBN 978-620-6-72840-5.

Publisher:
Sciencia Scripts
is a trademark of
Dodo Books Indian Ocean Ltd. and OmniScriptum S.R.L publishing group

120 High Road, East Finchley, London, N2 9ED, United Kingdom
Str. Armeneasca 28/1, office 1, Chisinau MD-2012, Republic of Moldova, Europe
Managing Directors: Ieva Konstantinova, Victoria Ursu
info@omniscriptum.com

Printed at: see last page
ISBN: 978-620-8-37861-5

PORT OVERCAPACITY AND UNDERACTIVITY

Detailed plan :

Introduction

- **Subject presentation**

  - Background to the growth of port infrastructures worldwide.

  - Importance of ports in the global and regional economy.

- **Objectives of the work**

  - Analyze the challenges and opportunities of overcapacity and underactivity.

  - Present sustainable solutions for the efficient management of port infrastructures.

- Methodology

  - Analytical approach with case studies, figures and global trends.

I. Background and importance of the ports

1. **Strategic role in international trade**.

   - Importance of ports in the global logistics chain.

   - Ports as distribution and transshipment hubs.

2. **Economic impact on surrounding regions**.

   - Impact on local employment and investment.

   - Direct and indirect economic benefits for the regions.

II. Definition of Concepts

1. **Overcapacity

   - Description: excess port infrastructure in relation to actual demand.

   - Types: unprofitable port infrastructures, unused storage areas.

2. **Sub-activity**

- Definition: reduced use of existing port facilities.

- Common causes and impact.

III. Issues

- **How do overcapacity and underactivity affect the operational efficiency and profitability of ports?

Part I: Analysis of Port Overcapacity

Chapter 1: Causes of overcapacity

1. **Increased investment in infrastructure**.

- Public and private investment.

- Competitive race between regional ports.

2. **Trade growth and demand anticipation**.

- Unrealized expectations of increased trade flows.

- Forecasting errors and lack of adaptation.

3. **Technological evolution and equipment modernization**.

- The role of new technologies in port expansion.

- Inconsistency with real needs.

Chapter 2: Consequences of overcapacity

1. **High maintenance costs**.

- Fixed costs for maintenance of unused infrastructure.

- Maintenance and modernization costs.

2. **Decrease in profitability**.

- Effects on profit margins and port revenues.

3. **Environmental impact

- Pollution from unused infrastructure.

- Congestion and unnecessary congestion of port sites.

Chapter 3: Case studies

1. **Analysis of ports with overcapacity**.

- Examples: European, American, African and Asian ports.

2. **Comparison with ports that have successfully managed their capacity**.

- Success stories in capacity management.

Part II: Analysis of Port Sub-Activity

Chapter 4: Causes of Under-activity

1. **Changes in trade routes**.

- Redirection of trade flows.

- Impact of new routes, e.g. the Arctic route.

2. **Increased competition between ports**.

- Scope of regional and international competition.

3. **Logistical and accessibility issues**.

- Infrastructure and access constraints.

Chapter 5: Consequences of under-activity

1. **Loss of competitiveness**.

- Ripple effect on local trade.

2. **Impact on local employment and the regional economy**.

- Unemployment, falling local investment.

3. **Risk of closure or conversion**.

- Reduction of port activity and conversion of infrastructure.

Chapter 6: Case studies

1. **Analysis of ports suffering from under-activity**.

- Developing ports and isolated port areas.

2. **Initiatives taken to revitalize these ports**.

- Revitalization and rehabilitation projects.

Part III: Solutions and Improvement Strategies

Chapter 7: Infrastructure optimization

1. **Reconfiguration of existing installations**.

- Adaptation and modularity of infrastructures.

2. **Targeted investments to meet real demand**.

- Allocation of resources according to need.

Chapter 8: Collaboration and partnerships

1. **Synergies between regional ports**.

- Cooperation and complementarity between ports.

2. **Public-private partnerships**.

- Joint projects for port development.

Chapter 9: Technological innovation

1. **Adoption of advanced technologies**.

   - Digitization and automation of operations.

2. **Use of data and analytics**.

   - Demand forecasting and efficient management.

Chapter 10: Public policies and regulations

1. **Role of governments

   - Planning and regulation of the port sector.

2. **Tax incentives**.

   - Encouraging investment in under-used ports.

Part IV: Future prospects

Chapter 11: International trade trends

1. **New trade routes**.

   - Initiatives such as the Belt and Road.

2. **Changes in supply chains**.

   - Impact of geopolitical changes.

Chapter 12: Sustainability and the environment

1. **Integration of environmental concerns**.

   - Ecological strategies for port activities.

2. **Reduced carbon footprint**.

   - Sustainable practices to limit pollution.

Chapter 13: Technological developments

1. **Digitization of port operations**.

   - Implications of digital management.

2. **Automation and drones

   - Robotization for greater efficiency.

Conclusion

1. **Summary of issues** (in French)

   - The importance of balancing over-capacity and under-activity.

2. **Call to Action

   - Recommendations for decision-makers and investors.

Bibliography and Appendices

- **Bibliography** : List of sources used.

- Appendices**: Graphs, maps and figures.

Preface to the book :

Ports, pillars of the global economy, provide a crucial connection between nations and continents, linking markets and transporting essential goods. Today, however, they face growing challenges from overcapacity and under-activity, caused by global trade shifts, massive and sometimes inadequate investment, as well as competition and fluctuations in demand.

This book explores in depth the causes, consequences and solutions of port overcapacity and underactivity. Using case studies from across the continents, we examine concrete situations and illustrate how some ports are managing to balance their infrastructures, while others are still struggling to adjust.

This book is aimed at professionals, decision-makers and students wishing to understand the challenges of contemporary port management, as well as those in the maritime sector looking for innovative, sustainable solutions.

The hope is that this book will provide a better understanding of these dynamics and inspire more effective and resilient port management for the future.

Dedication

To all those involved in the maritime and port sector,

This work is dedicated to those who work tirelessly in the shadows of the quays and docks, who keep our ports alive, these veritable crossroads of world trade. You, dockers, managers, engineers, economists and decision-makers, whose commitment and expertise are essential to the smooth running of our port infrastructures. You are the architects of international trade that transcends borders and unites nations in a complex network of dependencies and exchanges.

To those who, every day, face the challenges of growing overcapacity and persistent under-activity, this dedication recognizes your resilience and ingenuity. Your efforts to optimize operations, minimize environmental impact and reinvigorate ports are the lifeblood of an ever-changing global economy.

To those who believe in the future of our marine infrastructure, who invest in innovation and sustainability, and who collaborate to create sustainable solutions, I dedicate this book. May your passion and determination inspire others to embrace this vital sector with the same fervor.

Finally, this dedication is also addressed to future generations. May they inherit ports that not only meet the economic needs of their time, but are also models of sustainability, efficiency and innovation. May the challenges we face today be the springboards for a better future, where every port will be a beacon of prosperity for its communities and for the whole world.

With respect and gratitude,

MOUNCHEROU WOUYIE Mamadou Nasir

Introduction

Introducing the subject

Ports play a strategic role in the global economy, as nerve centers for maritime transport and trade. However, in recent decades, the rapid development of port infrastructures has led to overcapacity and underactivity in some ports around the world.

Background to the growth of port infrastructures worldwide

The rapid expansion of port infrastructure is a response to the growth in international trade and rising sea freight volumes. With world seaborne trade increasing by 4% a year in recent decades, many countries have sought to modernize and expand their ports in anticipation of this growing demand. Massive expansion projects have been carried out, such as the extension of the port of Shanghai**, now one of the largest in the world, or the port of Los Angeles**, which has seen its capacity double in less than two decades to meet the needs of the American West Coast.

However, this expansion has not always kept pace with stable, balanced demand, leading to situations of overcapacity and under-activity. In some cases, anticipated economic growth failed to materialize, leaving infrastructures largely under-utilized. Ports such as **Busan** in South Korea or **Valencia** in Spain illustrate this phenomenon, where the ambitious expansions carried out in the 2000s now exceed actual needs in terms of cargo flows.

Importance of ports in the global and regional economy

Ports play a central role in regional and national economies as hubs of economic, logistical and industrial activity. For example, the port of Rotterdam** generates nearly 3.5% of the Netherlands' GDP and supports a vast network of jobs in the transport, logistics and handling sectors. In Asia, **Singapore** and **Hong Kong** illustrate how ports can be at the heart of the national economy, providing jobs, stimulating exports, and attracting foreign investment. In Africa, the port of **Djibouti** has become a strategic hub for intercontinental trade and foreign investment, demonstrating the importance of ports in emerging economies.

Overcapacity and under-activity, however, threaten these economic benefits. In oversized ports, operating and maintenance costs rise, while under-utilized infrastructures struggle to generate sufficient revenues. This problem, at the crossroads of economic, logistical and environmental issues, has become a major challenge for port managers.

Objectives of the project

The aim of this book is to explore the challenges and opportunities linked to the phenomena of overcapacity and underactivity in the port sector, by analyzing their causes, consequences and possible solutions.

Analyze the challenges and opportunities of overcapacity and underactivity

The main aim of this book is to understand how overcapacity and underactivity influence the operational efficiency and profitability of ports, while identifying management opportunities to optimize their operation. International case studies are used to examine concrete examples, such as the port of **Taranto** in Italy, which struggles with underactivity due to its poor logistics connectivity, or the port of

Shanghai, which has been able to effectively manage high demand without falling into overcapacity.

Presenting sustainable solutions for efficient port infrastructure management

The book also proposes practical and sustainable solutions to port management challenges, taking into account economic, social and environmental realities. Solutions include infrastructure reconfiguration, inter-port cooperation, and the adoption of advanced technologies to optimize the use of facilities. The aim is to present strategies that can be adapted to local contexts and global economic dynamics, while integrating environmental concerns.

Methodology

In order to tackle the problem in a comprehensive way, this book adopts an analytical approach based on case studies, figures and global trends. The methodology includes:

Analytical approach with case studies, figures and global trends

1. **Case studies**: An in-depth analysis of ports with contrasting realities will illustrate the diversity of situations, from overcapacity in the USA and Europe, to the challenges of under-activity in developing ports in Africa and Asia.

2. **Quantitative data**: Statistics and indicators on port capacity and activity will be presented to support the analyses. Data comes from reliable sources such as the World Bank, the

Organisation for Economic Co-operation and Development (OECD) and UNCTAD.

3. **Analysis of global trends**: The book will be based on a study of global trade trends and technological developments influencing port management. The impact of new trade routes, such as those developed under the Belt and Road Initiative, will also be taken into account.

This methodological framework will provide a comprehensive and relevant analysis, offering readers practical insights into the challenges of port overcapacity and underactivity.

I. Background and importance of the ports

In today's globalized world, ports play a central role in economic development and international trade. They play a crucial role as trade hubs, logistics centers and economic engines for the surrounding regions. This first part aims to explore the strategic role of ports in the global logistics chain, as well as their local and regional economic impact.

1. Strategic role in international trade

As the nerve centers of international trade, ports play an essential role in the global logistics chain. They are the main entry and exit points for goods, and ensure connectivity between continents. The majority of world trade - some 80% by volume and over 70% by value, according to the United Nations Conference on Trade and Development (UNCTAD) - transits by sea, underlining the strategic importance of ports in the global economy.

Importance of ports in the global logistics chain

Ports are at the heart of global supply chains, facilitating the flow of goods and raw materials between production areas and consumer markets. For example, the port of Shanghai** in China, which is the world's busiest in terms of container volume, will handle around 47 million TEUs (twenty-foot equivalent units) in 2021. This makes it a critical logistics infrastructure, enabling the efficient distribution of finished goods, materials and consumer goods destined for international markets.

Similarly, the port of Rotterdam**, Europe's leading port, recorded a total goods flow of over 469 million tonnes in 2020. Its strategic position and high-performance infrastructure place it at the heart of trade between Europe and the rest of the world, directly linking European markets with global suppliers. This type of "hub" port functions like a distribution center, where goods can be grouped together, temporarily stored and then redistributed, thus generating cost savings for companies.

Ports as distribution and transshipment hubs

Some ports act as transshipment hubs, where containers are unloaded from one vessel and reloaded onto another en route to a final destination. Ports like **Singapore** and **Colombo** in Sri Lanka take advantage of their strategic location to play a major role in redirecting international trade flows. Singapore, for example, handles almost 85% of its container traffic via transshipment, making it a key gateway for intra-Asian and global trade.

Transshipment ports offer advantages in terms of speed and reduced logistics costs. For example, the port of **Dubai** (Jebel Ali) is one of the largest transshipment centers in the Middle East, handling around 14 million TEUs a year. This type of port makes it possible to consolidate cargoes, thereby reducing transport costs through better optimization of flows. Shipping companies also enjoy a competitive advantage, as these ports facilitate inventory management and shorten delivery times.

2. Economic impact on surrounding regions

In addition to their role in the global logistics chain, ports have a direct and significant impact on local and regional economies. They contribute to job creation, attract investment, and promote industrial and commercial development in the regions where they are located.

Impact on local employment and investment

Ports are often the main employers in the coastal regions where they are located, generating numerous direct and indirect jobs. The Port of Los Angeles**, for example, supports around 1.6 million direct and indirect jobs in the United States, including over 370,000 in California alone. These jobs cover a wide range of occupations, from cargo handling to logistics and administrative management, as well as customs and security services.

Ports also act as catalysts for investment. The port of **Tanger Med** in Morocco is a striking example. Since its inauguration, it has attracted major multinationals such as **Renault** and **Siemens**, which have set up production and logistics facilities there. Not only has the port strengthened Morocco's industrial sector, it has also generated over 80,000 direct and indirect jobs in the region, significantly boosting local gross domestic product (GDP).

In developing countries, ports are often engines of economic growth. The port of **Mombasa** in Kenya, the largest in East Africa, supports the economies of several landlocked countries in the region, including Uganda and Rwanda. The modernization of its infrastructure and the extension of its terminals, financed in part by foreign investment, are helping to reduce logistics costs and improve the competitiveness of African products on the world market.

Direct and indirect economic benefits for the regions

The economic impact of ports goes beyond jobs and direct investment. Port infrastructures generate indirect benefits for local economies. For example, the presence of major ports can stimulate the development of industrial and logistics ecosystems around them. At the **Port of Antwerp** in Belgium, a logistics zone covering some 1,600 hectares has been developed, attracting companies specializing in warehousing, packaging and distribution. In 2021, the Port of Antwerp handled some 240 million tonnes of goods, generating significant revenues for the Belgian economy and contributing directly to the growth of local businesses.

Economic spin-offs are also felt in the real estate and services sectors, which benefit from increased port activity. In the vicinity of the port of **Pointe-Noire** in the Republic of Congo, the increase in maritime traffic has led to the creation of numerous local businesses, contributing to the diversification of the regional economy. In addition, support services such as repair workshops and catering services are flourishing around port facilities, contributing to the economic vitality of these areas.

The economic impact of ports is both vast and complex. By acting as levers of growth, ports create sustainable economic opportunities for the regions around them, while strengthening the integration of countries into the global economy. These essential infrastructures not only enable trade to take place, but also provide long-term socio-economic benefits for society as a whole.

II. Definition of Concepts

Before delving into the analysis of port-related issues, it is essential to define the key concepts that structure the current debate around port infrastructure management. The notions of "overcapacity" and "underactivity" refer to common issues in the port sector, linked to the adequacy between available facilities and the volumes of goods actually handled. Both phenomena present specific challenges for port managers, investors and regional authorities. This section examines the characteristics, types and impacts of each of these concepts, to shed light on their influence on port efficiency and profitability.

1. Overcapacity

Overcapacity in the port sector occurs when there is an excess of infrastructure in relation to the real needs of maritime trade. In other words, port facilities are under-utilized, generating high costs without a commensurate return on investment. This phenomenon is particularly noticeable in ports that invested heavily in infrastructure before a downturn or slowdown in trade.

Description: excess port infrastructure compared with actual demand

Overcapacity means that quays, terminals, storage areas or handling equipment remain unused or underused for a large part of the year. For example, in **Spain**, the creation of new ports and the extension of existing infrastructures in the 2000s generated excess capacity, particularly in the ports of **Valencia** and **Algeciras**. From 2010 onwards, demand for maritime traffic slowed, leaving a significant part of the infrastructure unused, resulting in maintenance expenditure without sufficient financial returns.

This phenomenon of overcapacity is not just a question of surface area, but also of equipment and manpower mobilized without adequate return. The annual cost of under-utilized infrastructure is estimated at several million euros, burdening local finances and holding back investment in other, more profitable sectors.

Types: unprofitable port infrastructure, unused storage areas

Overcapacity can take several forms in the port context:

- **Unprofitable infrastructure**: Some ports, anticipating continued traffic growth, have built excess quays and terminals. These unused infrastructures, such as certain terminals in the port of Rotterdam**, still require maintenance and personnel costs, creating an economic imbalance.

- **Unused storage areas**: In the absence of sustained traffic, storage areas remain vacant, as in the ports of **Malaysia**, where a massive expansion took place in the 2010s. Container flow forecasts have not always materialized, leaving parts of the storage areas unoccupied, generating unnecessary management costs.

This overcapacity is sometimes due to investment decisions based on optimistic growth projections. However, when demand stagnates or declines, ports find themselves with surplus infrastructure that is difficult to make profitable, resulting in increased financial pressure on port authorities.

2. Sub-activity

Under-activity, on the other hand, is a situation where the available port infrastructures are little used. Unlike overcapacity, which refers to an excess of physical capacity, underactivity refers to a low level of utilization of existing facilities, even if capacity appears to be adjusted to potential needs.

Definition: reduced use of existing port facilities

Under-activity manifests itself in minimal dock occupancy, low vessel frequency, and a reduced number of handling operations. It can be caused by economic, logistical or competitive factors. For example, the port of Lagos in Nigeria experiences periods of under-activity due to frequent road congestion and inefficiencies in the local logistics chain, discouraging shipping lines from docking there regularly.

Common causes and impact

There are several possible reasons for a port's under-activity:

- **Economic factors** : Economic recessions or financial crises reduce the volume of maritime trade, as observed during the 2008 financial crisis, when several ports experienced a marked drop in activity. At the time, ports in **Greece** and **Spain** were particularly hard hit, with a significant reduction in cargo traffic.

- **Logistics factors** : Transport infrastructure and inefficient logistics management can limit a port's attractiveness. For example, the port of **Cotonou** in Benin has lost market share to its neighbors due to its logistical challenges, including frequent delays in clearing goods through customs.

- **Increased competition** : Geographical proximity to other, more efficient or better-equipped ports can also lead to under-activity. The port of **Hamburg**, although one of Germany's largest, has had to cope with a decline in traffic to the benefit of Rotterdam, which benefits from more modern land transport infrastructures.

Under-activity has considerable economic impacts: the fixed costs of operating a port do not decrease with the volume of activity, which means that even in the absence of cargo flows, significant expenditure on personnel, maintenance and security continue to strain the port's budget. What's more, the surrounding regions also suffer the effects of this under-utilization of facilities, losing out on the potential revenues and indirect benefits associated with dynamic port activity.

The concepts of overcapacity and underactivity, while often appearing in different contexts, pose similar challenges for port managers. Excess capacity leads to unprofitable costs, while under-activity reduces a port's competitiveness against better-positioned rivals.

III. Issues

The issue of efficient port infrastructure management is crucial in a context where international competition and fluctuations in maritime trade dictate the economic dynamics of many regions. While many ports around the world continue to invest in additional capacity, suboptimal utilization of these facilities, whether due to over-capacity or under-activity, raises significant operational and financial stakes. Both issues can have a major impact on the profitability of ports, the attractiveness of their services, and their regional impact.

Key question: How do overcapacity and underactivity affect the operational efficiency and profitability of ports?

There are several key aspects to this question:

- **Operational efficiency**: A port in a situation of overcapacity or under-activity often faces challenges in ensuring optimal use of its resources, such as personnel, equipment and infrastructure. Under-utilization can also lead to a deterioration in the quality of services offered, as investments are not maximized, and resources remain tied up.

- **Profitability** : Managing a port represents a high cost, in terms of infrastructure maintenance, salaries, security and logistics. In a context of overcapacity, these costs are not offset by the revenues generated, resulting in an operating deficit. For example, Asian ports such as **Kaohsiung** in Taiwan find themselves in a situation of reduced profitability due to the emergence of regional competitors who are capturing a growing share of the market.

The impact of these phenomena goes beyond the financial sphere: a port that does not reach its full potential becomes less attractive to investors, shipping lines and local businesses. Managing overcapacity and underactivity is therefore a strategic challenge for the future of the port sector, which must respond to fluctuations in international trade while preserving its efficiency and economic viability.

Part I: Analysis of Port Overcapacity

With maritime trade accounting for the majority of international trade, many ports find themselves in a situation of overcapacity. This phenomenon, while driven by the desire to improve competitiveness and meet growing demand, can lead to inefficiencies and wasted resources. This first part explores the root causes of overcapacity, analyzing economic motivations, forecasting errors, and the impact of modern technologies on port infrastructure expansion.

Chapter 1: Causes of overcapacity

1. Increased investment in infrastructure

One of the main factors behind overcapacity is the massive increase in both public and private investment in port infrastructure. These investments, driven by the desire to modernize and expand ports, have been amplified by growing competition between regions. Governments and private investors see ports as a lever for economic development, encouraging the construction of additional terminals and the modernization of equipment.

- **Public and private investment**: In Europe, for example, projects financed by European funds have enabled the expansion of several ports, such as the port of **Valencia** in Spain. In the USA, the Port of Los Angeles** has also seen a flow of private and public investment to improve its facilities, but this has created a situation where available capacity is not always used to its full potential.

- **Competition between regional ports** : Asian ports are a good illustration of this dynamic. Competition between the ports of **Singapore**, **Shenzhen** and **Shanghai**, each seeking to become the regional logistics hub, has led to massive expansions, increasing capacity far beyond current demand. This one-upmanship has often led to under-utilized infrastructures, creating a situation of overcapacity.

2. Trade growth and demand anticipation

Another significant cause of overcapacity is linked to unrealized expectations of increased trade flows. World trade growth projections, while sound, have sometimes proved over-optimistic, leading to over-investment.

- **Unrealized gains from increased trade flows**: In the 2000s, strong growth in world trade led to a series of port expansion projects. However, the 2008 financial crisis led to an unforeseen slowdown in trade, leaving many ports with unused infrastructure. For example, the **port of Pusan** in South Korea, having anticipated a surge in trade with China, found itself with excess capacity when regional trade stagnated.

- **Forecasting errors and lack of adaptation**: Forecasting errors are common, as they are based on complex economic growth models. In some cases, ports have failed to adapt their investment strategy. The **port of Taranto** in Italy, for example, had anticipated growth in trade with Eastern Europe and the Middle East, but suffered a drop in activity, illustrating the limits of a strategy based on uncertain forecasts.

3. Technological evolution and equipment modernization

Modernization of port infrastructures, while beneficial, can also lead to overcapacity when new technologies are not fully compatible with real needs or local practices. The adoption of advanced technologies is often accompanied by a physical expansion of facilities to accommodate new equipment.

- **The role of new technologies in port expansion**: The automation and digitization of port processes often requires larger facilities to handle automated systems. At the **port of Rotterdam**, for example, automation led to the creation of additional terminals, but the technological transition took longer than expected, and the additional facilities were not fully utilized at first.

- **Incompatibility with real needs** : Technological advances, while essential for efficiency, can sometimes prove unsuited to market needs. The Hambantota **port in Sri Lanka, built with modern equipment in anticipation of increased trade in South Asia, found itself under-utilized as regional trade flows failed to meet expectations, illustrating the impact of investments misaligned with real needs.

This chapter highlights the fact that port overcapacity is the result of a combination of economic, forecasting and technological factors, illustrating the need for more flexible planning that is better adapted to the changing dynamics of world trade.

Chapter 2: Consequences of overcapacity

Port overcapacity has detrimental effects both financially and environmentally, limiting profitability and creating pressure on local resources. This chapter explores the various consequences of overcapacity, detailing the high costs associated with maintenance, the reduced profit margins for port operators, and the environmental impacts generated by under-utilized infrastructure. These elements show that overcapacity is not simply a question of space availability, but a source of economic and ecological concern.

1. High maintenance costs

Port infrastructures require substantial fixed maintenance costs, even if they are not operated at full capacity. These costs include maintenance of quays, loading and unloading equipment, and periodic modernization to avoid obsolescence.

- **Fixed charges for maintenance of unused infrastructure** : Ports must maintain secure, operational facilities, whether or not they are in full use. For example, at the **port of Cagliari** in Italy, the costs of maintaining unused infrastructure, following the reduction in flows, have had a considerable financial impact on the region's budget, with maintenance expenses amounting to several million euros a year.

- **Maintenance and modernization costs**: Equipment modernization, while crucial to maintaining competitiveness, imposes additional costs on ports with overcapacity. In the **United States**, some West Coast ports have seen their modernization costs rise by 15% to 20% due to under-utilization of certain areas, which require regular work to avoid deterioration. These recurring costs weigh on the overall profitability of port infrastructures.

2. Reduced profitability

Overcapacity directly affects port profitability, as under-utilized facilities do not generate sufficient revenues to cover fixed and operating costs. Profit margins are shrinking, and some ports are struggling to achieve stable revenue levels.

- **Effects on port profit margins and revenues** : Ports that do not use their infrastructure to full capacity are seeing a decline in profitability. At the **port of Qingdao** in China, rapid infrastructure expansion has led to chronic overcapacity, reducing profit margins by 8% in recent years. Revenues, though growing, are not offsetting the rising costs of managing this unused space.

This decline in profitability is forcing many ports to reduce their tariffs to attract more traffic, which can lead to a race to the bottom and a reduction in service quality. Some ports in Southeast Asia, such as **Port Klang** in Malaysia, have had to offer discounts and tax breaks to attract shipping lines, affecting their long-term profitability.

3. Environmental impact

Overcapacity is not just a financial challenge; it also has a significant environmental impact. Unused or partially exploited port infrastructures contribute to environmental degradation through pollution and congestion on port sites.

- **Pollution from unused infrastructure** : Overcapacity ports generate waste and pollutants even in the absence of heavy activity. For example, under-utilized facilities continue to emit greenhouse gases linked to equipment maintenance and the operation of port facilities, contributing to local air pollution. In **Europe**, ports such as **Genoa** in Italy have observed an increase in pollution due to equipment in operation, even in periods of low traffic.

- **Unnecessary congestion and overcrowding of port sites**: In some overcapacity ports, unused areas take up space that could be reallocated, contributing to unnecessary congestion. The Port of Rotterdam** in the Netherlands, despite its size, is under increased environmental pressure due to partially unused facilities, and the site has had to implement initiatives to minimize the impact of this congestion, such as installing sensors to better manage the influx of ships and reduce local pollution.

These examples show that overcapacity is not just an economic problem, but also an environmental challenge, requiring ports to review their planning and management strategies to minimize negative effects.

Chapter 3: Case studies

To better understand the dynamics of port overcapacity, this chapter analyzes concrete examples of ports in different regions of the world and examines their varied approaches to capacity management. It compares ports with persistent overcapacity with those that have succeeded in optimizing infrastructure utilization. This international overview identifies best practices and innovative strategies that could inspire more effective port capacity management.

1. Analysis of ports with overcapacity

In several regions of the world, ports are experiencing situations of overcapacity that are impacting on their profitability and performance. These examples illustrate the causes of overcapacity, the specific challenges and the lessons learned from these experiences:

- **Europe**: The port of **Algesiras** in Spain was designed to support strong growth in demand for sea transport between Europe and Africa, but has not achieved the volumes hoped for due to competition from neighboring ports such as **Tanger Med** in Morocco. At around 60% capacity utilization, Algesiras is incurring high maintenance costs without generating the revenues initially anticipated, illustrating the risks of over-anticipating demand.

- **United States**: The port of **Savannah** in Georgia has invested heavily in modern infrastructure to become a key hub for US imports and exports. However, unforeseen fluctuations in demand and the growth of competing ports have reduced the use of certain facilities. As a result, Savannah must now redirect part of its infrastructure towards logistics uses to optimize occupancy and minimize maintenance costs.

- **Africa**: The port of **Lagos** in Nigeria has seen its capacity grow to meet regional demand. However, management and road infrastructure problems limit the fluidity of operations, and part of the facilities remain under-utilized. This situation shows that overcapacity is not only the result of demand forecasts, but also of inadequate management of logistics infrastructures.

- **Asia**: The port of **Busan** in South Korea, long the leader in Asia, is facing overcapacity due to the rise of Chinese ports such as **Shanghai** and **Ningbo**. Busan has tried to diversify its services by offering logistics solutions and developing industrial zones to attract businesses, but competition remains intense, and part of the facilities are under-utilized.

2. Comparison with ports that have successfully managed their capacity

Some ports have adopted effective strategies to adjust their capacity in line with demand and avoid overcapacity. By analyzing these cases, we can identify practices that promote sustainable management of port capacity:

- **Port of Rotterdam, Netherlands**: Renowned for its innovative approach, the Port of Rotterdam has invested in flow management and space optimization technologies. Thanks to an advanced planning system, it adjusts the capacity of its quays and storage areas according to demand forecasts. In periods of low activity, certain facilities are converted for logistics use, reducing the risk of overcapacity while generating additional revenue.

- **Port of Singapore**: As a global hub, the Port of Singapore optimizes its capacity through integrated logistics and automation. It uses artificial intelligence tools to predict capacity requirements and adjust space allocation accordingly. For example, in the event of a drop in demand, it redirects activities to warehouse or transshipment zones. This flexibility has enabled Singapore to maintain constant utilization of its infrastructures.

- **Tanger Med Port, Morocco**: Faced with rapid growth, Tanger Med has opted for phased development to avoid overcapacity. Each extension is built according to actual demand, which has enabled the port to maximize the occupancy of its facilities. By using a modular strategy, Tanger Med is able to respond to fluctuations in demand while minimizing the maintenance costs associated with potential overcapacity.

These case studies demonstrate that overcapacity can be effectively managed through a combination of technology, organizational flexibility and adaptive planning. Ports that anticipate not only growth, but also variations in demand, are better prepared to avoid the financial and environmental consequences of overcapacity.

Part II: Analysis of Port Sub-Activity

This second part explores port under-activity, a common problem affecting the profitability and performance of port infrastructures. Under-activity occurs when available facilities are insufficiently utilized, often due to changes in trade routes, increased competition or logistical and accessibility problems. This section examines the major causes of under-activity, drawing on concrete examples, to better understand how ports can anticipate and mitigate these situations.

Chapter 4: Causes of Under-activity

Under-activity in ports stems from multiple factors, some of which are intrinsic to global changes and developments in the shipping industry. The following causes highlight the pressures exerted by changes in trade routes, competition between ports and logistical obstacles.

1. Changes in trade routes

Trade routes are constantly evolving in response to new economic opportunities, technological advances and geopolitical pressures. This can have a direct impact on port activity:

- Redirecting trade flows**: For example, the introduction of new routes such as the Arctic route, which links Asia and Europe more quickly by bypassing traditional channels, reduces dependence on Mediterranean ports such as Suez or Piraeus. Melting ice in the Arctic has opened up a new, shorter and more competitive transport route. Ports that are heavily dependent on certain flows risk a drop in activity if these flows are redirected to alternative routes.

- Geopolitical changes**: Geopolitical tensions can also influence the reorganization of trade routes. For example, tensions between the USA and China have led some companies to seek alternative sources of supply, reducing activity at ports specializing in Chinese goods. This is reflected in US ports such as Los Angeles, where a drop in imports from China has led to temporary under-activity.

2. Increased competition between ports

Increased competition is another major cause of port underactivity, pushing some ports to lose market share to more attractive competitors:

- Scope of regional and international competition**: In regions where ports are numerous and close to each other, competition is intense. In Asia, for example, the ports of Singapore, Busan and Hong Kong all compete to attract major ocean carriers. This competition can lead the least efficient or least modernized ports to lose customers, resulting in prolonged under-activity.

- The case of the Port of Durban in South Africa**: Durban, long considered the leading port in Southern Africa, has recently seen its activity reduced due to growing competition from emerging ports such as Walvis Bay in Namibia, which offers modern infrastructures and reduced handling times. This competition has prompted Durban to review its strategy and invest in modernizing its facilities to regain market share.

3. Logistics and accessibility

Logistical challenges and accessibility obstacles often limit the ability of ports to make full use of their infrastructure:

- Infrastructure constraints**: Some ports suffer from bottlenecks in their logistics chain, reducing the efficiency of their operations. For example, the port of Lagos in Nigeria suffers from road accessibility problems, making it difficult to transport goods to domestic markets. This logistical limitation results in under-utilization of the infrastructure, despite significant potential demand.

- Environmental conditions and regulatory restrictions**: Some ports also have to cope with specific environmental conditions, such as tides or currents, which limit their ability to accommodate large vessels. In addition, strict regulations or government restrictions can sometimes slow down port operations, affecting infrastructure utilization. One example is the port of Hamburg in Germany, where environmental constraints to protect the Elbe prevent sufficient dredging of the river, making it difficult for large ships to access the port and thus reducing port activity.

Under-activity in ports is often the result of external changes, increased competition and internal logistical challenges. For ports faced with this situation, rapid adaptation and upgrading of their assets are essential to attract new business and improve profitability.

Chapter 5: Consequences of Sub-Activity

The under-activity of ports has profound repercussions on their profitability, regional competitiveness and local economies. This chapter explores these consequences in detail, showing how the under-utilization of port infrastructure can lead to a loss of competitiveness, a reduction in local employment, and even risks of closure or conversion of port facilities.

1. Loss of competitiveness

Under-activity leads to a significant drop in competitiveness for the ports concerned, which see their attractiveness reduced in the face of more dynamic competitors:

- Ripple effect on local trade** : When a port is under-utilized, shipping lines and logistics operators often turn to busier, more efficient alternatives. This leakage of commercial activity not only weakens the port, but also the entire economic network that depends on its smooth operation, from nearby industrial zones to logistics service providers. In North America, for example, medium-sized ports such as Baltimore face competition from large gateway ports such as New York and Savannah, which attract more traffic thanks to their higher business volumes and modernized infrastructures.

- **Loss of expansion opportunities**: An under-active port lacks the revenue to reinvest in its infrastructure, creating a vicious circle in which its lack of competitiveness is progressively reinforced. Examples from Asia illustrate this phenomenon, notably in Malaysia's secondary ports, which, faced with the growth of Singapore and Port Klang, are struggling to maintain sufficient activity to remain competitive.

2. Impact on local employment and the regional economy

Under-activity also has direct socio-economic consequences for the surrounding regions, particularly in terms of employment and investment:

- Unemployment and falling local investment** : When port activity declines, directly related jobs such as dockers and stevedores are immediately impacted. At the same time, indirect jobs in the supply, logistics and transport sectors also suffer. A notable example is the port of Marseille, which, in a period of under-activity, has seen thousands of jobs in the logistics sector affected, resulting in lower revenues for the Provence-Alpes-Côte d'Azur region.

- **Reduction in local tax revenues** : Declining activity in a port leads to a reduction in the tax revenues collected by local authorities, impacting the resources available for regional economic development. For example, in Spain, secondary ports faced with under-activity, such as Huelva and Tarragona, have seen a drop in tax revenues, compromising budgets earmarked for infrastructure and public services.

3. Closure or conversion risks

For some ports, prolonged under-activity may lead to extreme scenarios involving the closure or conversion of their facilities:

- Reduced port activity**: In the absence of an effective recovery, some ports end up limiting their operations or reducing their capacity by closing certain facilities. For example, the Port of Valencia in Spain had to restructure its operations by closing terminals to concentrate on the most profitable ones.

- **Infrastructure conversion**: Some ports faced with under-activity are choosing to diversify their facilities by turning to other activities to optimize infrastructure use. In the United States, the Port of Oakland, faced with prolonged underactivity, has reoriented some of its areas for residential and commercial development. In Europe, the Port of Rotterdam has transformed inactive port facilities into zones dedicated to renewable energies, such as wind farms and solar installations.

In short, under-activity is a major challenge for ports, threatening their competitiveness, their contribution to the regional economy, and sometimes even their survival as logistics hubs. These consequences underline the importance for ports to adapt to changes in world trade, remain competitive and develop diversification strategies to avoid under-utilization of their facilities.

Chapter 6: Case studies

In this chapter, we will look at concrete examples of ports suffering from under-activity, exploring the factors that have contributed to their decline and the measures being implemented to revive their activity. Drawing on specific case studies, we will analyze the unique challenges these ports face, and the initiatives being taken to breathe new life into these crucial infrastructures.

1. Analysis of ports suffering from under-activity

Some ports, particularly those in remote or developing regions, struggle to attract sufficient business to ensure their profitability. These ports are often in competition with larger, more modern facilities, or are affected by changes in trade routes:

- Example of the port of Djibouti**: Although important for East Africa, the port of Djibouti has had to cope with a decline in activity linked to the development of the port of Berbera, in Somalia, which is attracting a growing share of regional traffic. Competition between these two ports has led Djibouti to review its strategies to boost its attractiveness and diversify its logistics offering.

- Port of Cotonou, Benin**: In West Africa, the Port of Cotonou has also suffered a period of under-activity due to increased competition from neighboring ports, notably Lomé and Tema. This decline is partly attributable to infrastructure limitations and logistical management problems that have made the port less attractive to maritime trade operators.

- Port of Rio Grande, Brazil**: Although strategic for the south of Brazil, the port of Rio Grande has encountered difficulties in capturing major traffic flows due to its distance from the country's major industrial centers, such as São Paulo. Distance and lack of connectivity with rail and road networks have limited its development potential.

2. Initiatives to revitalize these ports

To combat under-activity, several ports have implemented revitalization initiatives targeting infrastructure improvements, diversification of activities and development of new trade routes. Examples of successful projects include

- Revitalization at the Port of Valencia, Spain** : Faced with a decline in activity, the Port of Valencia launched a vast program to modernize its facilities, focusing on digitalization and logistics optimization. By integrating innovative port management systems, Valencia has succeeded in attracting more operators and increasing the fluidity of its operations, which has contributed to a significant increase in traffic.

- Mombasa Port Development Project, Kenya**: To meet the challenges of under-activity and regional competition, the Port of Mombasa has invested in the construction of an additional container terminal and in improving rail connectivity with Nairobi and other inland regions. This initiative aims to make Mombasa more attractive to operators and increase the volume of goods transiting through the port.

- **Transformation of the port of Gdansk, Poland** : Faced with periods of under-activity in the 1990s, the port of Gdansk set up a reconversion strategy, positioning itself as a hub for Central and Eastern European trade. With the support of the European Union and private investment, the port has modernized its facilities and diversified its services to include intermodal transport and green logistics, thereby strengthening its competitive position.

In conclusion, these case studies show how well thought-out and strategic initiatives enable ports to overcome underactivity. Through targeted investment and adaptation to global trade trends, ports can reposition themselves to once again become dynamic and attractive players in the logistics and international trade sector.

Part III: Solutions and Improvement Strategies

In a context where overcapacity and underactivity of port infrastructures pose increasing challenges to the operational efficiency and profitability of ports, it becomes crucial to explore solutions and improvement strategies. This section aims to identify and analyze innovative and pragmatic approaches that can be implemented to optimize the use of port facilities, ensure their long-term viability and meet the needs of ever-changing global trade.

We'll start by examining the need to reconfigure existing infrastructures to adapt them to contemporary requirements, taking into account technological advances and industry best practices. Then we'll look at the importance of targeted investment, enabling ports to respond effectively to actual demand while anticipating future trends.

By integrating figures, exemplary case studies and sectoral perspectives, this section will highlight concrete strategies that have proved successful in various ports around the world. The aim is to provide decision-makers and port managers with practical tools and recommendations for navigating an increasingly complex maritime environment. The transformation of port infrastructures, through innovative and sustainable solutions, is not only an opportunity for improvement, but also a necessity if ports are to remain competitive on a global scale.

Chapter 7: Infrastructure optimization

In this chapter, we will focus on solutions and strategies for optimizing port infrastructures to meet the challenges of overcapacity and underactivity. By focusing on the reconfiguration of existing facilities and targeted investments, we will explore how ports can improve their operational efficiency, profitability and sustainability. Concrete examples will illustrate the importance of adapting port infrastructures to changes in global trade.

1. Reconfiguration of existing installations

To meet today's challenges, many ports need to rethink and reconfigure their existing facilities. This means making infrastructures more modular and adaptable to fluctuations in demand. Here are a few approaches:

- **Terminal flexibility**: Modular terminals enable operations to be rapidly adapted to changing market needs. For example, Rotterdam's container terminal has integrated multi-purpose zones that can be used for different types of cargo, enabling a rapid response to variations in volume. According to a report by the International Maritime Organization (IMO), this approach has led to a 15% increase in operational efficiency.

- **Use of advanced technologies**: The integration of technologies such as the Internet of Things (IoT) and artificial intelligence (AI) enables better resource management. For example, the Port of Singapore uses sensors to monitor infrastructure utilization in real time and optimize ship loading and unloading. This technology has reduced ship waiting times by 20%, improving the port's overall efficiency.

- **Dynamic storage spaces**: The creation of dynamic storage spaces, capable of changing assignment according to seasonal needs or changes in the supply chain, is essential. The Port of Los Angeles has developed flexible storage areas that adapt to the types of goods handled, maximizing space utilization while reducing operating costs.

2. Targeted investment to meet real demand

Another crucial aspect of improving port performance is the strategic allocation of investments according to real market needs. Here's how it works:

- **Market trend analysis**: Before investing in new infrastructure, ports need to carry out in-depth market studies to understand current and future trade trends. For example, the Port of Hamburg used predictive models to anticipate changes in demand, enabling it to allocate resources for the expansion of its container terminal only when justified by the data.

- **Public-private partnerships** : Targeted investments can also be made through public- private partnerships (PPPs), which enable financial risks to be shared and project development to be accelerated. The project to extend the port of Cochin in India, financed by a collaboration between the government and private investors, has modernized the infrastructure to meet the specific needs of local and regional maritime trade.

- **Sustainability support**: Ports must also invest in sustainable infrastructure to meet growing expectations of environmental responsibility. For example, the Port of Barcelona has integrated charging facilities for electric ships and modern waste management systems, which not only meet environmental demands, but also enhance the port's attractiveness to carbon-conscious shipping companies.

In short, optimizing port infrastructures requires intelligent reconfiguration and targeted investments that meet real market requirements. Ports that adopt these strategies can not only improve their operational efficiency, but also guarantee their competitiveness in the global marketplace. The use of advanced technologies and the implementation of strategic partnerships play a crucial role in this transformation, enabling ports to adapt to the challenges of the 21st century.

Chapter 8: Collaboration and partnerships

Collaboration and partnerships are key to overcoming the challenges of port overcapacity and underactivity. As economic and business environments become increasingly complex, cooperation between port players, whether public or private, can offer innovative solutions to improve operational efficiency and profitability. This chapter will explore synergies between regional ports and the importance of public-private partnerships in port development.

1. Synergies between regional ports

Regional ports can benefit from increased cooperation to optimize their operations and better meet the needs of maritime trade. For example, collaboration between neighboring ports can share resources, harmonize customs procedures and improve multimodal logistics. A study by the International Maritime Organization (IMO) has shown that ports that establish cooperative networks see a 20% increase in operational efficiency.

Synergies can also take the form of joint projects, such as the development of transshipment zones or shared logistics facilities. A concrete example is the cooperation between the Port of Rotterdam and the Port of Antwerp, which have set up initiatives to facilitate the transfer of goods and improve the supply chain in Europe. These collaborations not only reduce operating costs, but also enhance the competitiveness of the ports on the international stage.

2. Public-private partnerships

Public-private partnerships (PPPs) represent a strategic approach to port infrastructure development. By pooling the resources and expertise of the public and private sectors, these partnerships make it possible to invest in large-scale projects while sharing the financial and operational risks. According to a World Bank study, port projects carried out in partnership with private companies have a 30% higher probability of success than those financed solely by public funds.

A significant example is the development of the Port of Hamburg, where collaboration between the local government and private companies has modernized infrastructure and increased container handling capacity. This initiative led to a 15% increase in port traffic volume in just three years, illustrating the effectiveness of PPPs in the port sector.

Furthermore, collaboration and partnerships are crucial levers for improving port performance. By fostering synergies between regional ports and engaging in public-private partnerships, it is possible to optimize infrastructure utilization, increase competitiveness and ensure the sustainable development of the port sector. These initiatives represent opportunities for ports to adapt to contemporary challenges while meeting the growing demands of international trade.

Chapter 9: Technological innovation

Technological innovation is a key factor in improving port efficiency in the face of overcapacity and underactivity. By integrating advanced technological solutions, ports can not only optimize their operations, but also adapt to rapid changes in the global marketplace. This chapter focuses on the adoption of advanced technologies, such as digitization and automation, as well as the use of data and analytics for demand forecasting and efficient management of port resources.

1. Adoption of advanced technologies

Digitization and automation have become priorities for ports wishing to increase their operational efficiency. Implementing digital systems streamlines processes, reduces waiting times and improves goods traceability. For example, the Port of Rotterdam has invested heavily in digital technologies, resulting in a 30% reduction in container processing times.

The automation of port operations, notably through the use of autonomous vehicles and automated cranes, increases productivity while minimizing human error. According to a McKinsey study, ports that adopt automated systems can improve their processing capacity by 50%, while reducing operating costs by 15%.

An illustrative example is the port of Qingdao in China, which has implemented an automated control system for cargo management, enabling a significant reduction in labor costs and an increase in efficiency. The integration of advanced technologies is helping to transform the port into a modern hub, capable of meeting the needs of international trade in an agile and efficient manner.

2. Use of data and analytics

The use of data and analytics plays a crucial role in modern port management. Thanks to advanced analytical tools, ports can forecast demand, analyze trends and optimize the use of their infrastructures. Real-time data analysis also enables them to anticipate congestion and better manage traffic flows.

Ports can take advantage of Big Data to improve their strategic planning. For example, the Port of Los Angeles uses machine learning algorithms to analyze sea and land traffic data, enabling it to optimize ship arrival and departure times. This approach has led to a 25% reduction in waiting times for ships, contributing to greater profitability.

In addition, the integration of IoT (Internet of Things) systems enables real-time monitoring of port operations. This helps detect anomalies and implement corrective measures quickly, improving overall operational efficiency. Studies show that the use of advanced analytics in ports can increase the accuracy of demand forecasts by 20%.

Ultimately, technological innovation, through the adoption of advanced technologies and the use of data and analytics, represents an essential solution for ports facing the challenges of overcapacity and underactivity. By investing in these areas, ports can not only improve their operational efficiency, but also prepare for the future demands of international trade. These technological transformations are not only operationally beneficial, but also strengthen the long-term sustainability and competitiveness of ports.

Chapter 10: Public policies and regulations

Public policies and regulations play a crucial role in the management of port infrastructures, especially in the face of overcapacity and underactivity. A strategic approach on the part of governments can promote the harmonious development of ports and ensure their competitiveness in the global marketplace. This chapter looks at the role of governments in planning and regulating the port sector, as well as the fiscal incentives in place to encourage investment in under-utilized ports.

1. The role of governments

Governments are responsible for defining a strategic framework for port infrastructure development. This includes long-term port planning, taking into account international trade trends, economic fluctuations and technological developments. For example, the European Union has developed transport policies that aim to enhance port connectivity while improving operational efficiency. In 2021, a European Commission report stressed the importance of integrated planning to meet the challenges of port congestion and environmental impact.

Governments must also regulate the port sector to ensure a balance between competition and cooperation between ports. This involves establishing clear rules on access to port infrastructure, tariffs and services. In Australia, for example, the New South Wales Ports Regulatory Authority has established a regulatory framework to ensure cost transparency and competitive port services.

In addition, governments can encourage collaboration between ports and other players in the logistics chain, such as overland transport companies, customs agencies and freight operators. This collaborative approach can optimize the use of existing infrastructure and reduce the risk of under-activity.

2. Tax incentives

Tax incentives are an important lever for attracting investment to under-utilized ports. Governments can offer tax breaks, subsidies and tax credits to encourage port infrastructure development and service improvements. For example, under the Port Infrastructure Development Program in the USA, the federal government has allocated funds to support port modernization projects, with a particular focus on under-utilized regional ports.

In France, the "Loi d'Orientation des Mobilités" (LOM) law introduces incentives for ports wishing to improve their competitiveness and attract private investment. These include tax exemptions for infrastructure projects that promote multimodal transport and reduce the carbon footprint of port operations.

Governments can also introduce investment policies that promote the sustainable development of ports. For example, by setting up subsidy programs for the installation of eco-responsible equipment, they encourage not only the modernization of infrastructures, but also the reduction of the environmental impact of port activities.

This being the case, public policy and regulation are essential for the efficient management of ports in the face of the challenges of overcapacity and under-activity. The proactive role of governments in planning, regulating and offering tax incentives can help revitalize under-utilized ports and optimize existing infrastructure. These measures not only promote the competitiveness of ports, but also their adaptation to the future requirements of international trade and environmental sustainability.

Part IV: Outlook

The future of port infrastructures is closely linked to developments in international trade, new geopolitical dynamics and technological advances. This chapter looks at the major trends shaping the global port landscape, including new trade routes and evolving supply chains. By examining these trends, we can better understand how ports can adapt and prepare for future challenges and opportunities.

Chapter 11: International trade trends

In an ever-changing world, ports need to anticipate the changes affecting international trade. New trade routes, such as the Belt and Road Initiative (BRI), and evolving supply chains directly influence the way ports operate and organize themselves.

1. New trade routes

The Belt and Road initiative, launched by China in 2013, aims to create a transport and trade network linking Asia, Europe and beyond. This colossal project includes investment in port, rail and road infrastructure, facilitating trade between participating countries. According to a report by the Asian Development Bank, around $1,300 billion is expected to be invested in this initiative by 2030, creating new opportunities for ports worldwide.

Ports located along these new trade routes are strategically positioned to benefit from increased trade flows. For example, the port of Gwadar in Pakistan, developed as part of the BRI, has become a major hub for trade between China and the Middle East. However, this new dynamic also poses challenges, including increased competition between ports and the need to improve infrastructure to handle rising traffic volumes.

2. Supply chain evolution

Global supply chains are evolving rapidly in response to factors such as geopolitical shifts, economic crises and environmental concerns. Trade tensions between major powers, such as the USA and China, have led many companies to reconsider their production and distribution strategies. For example, the concept of "decoupling" has gained in popularity, prompting some companies to relocate production closer to their target markets to reduce their dependence on long and vulnerable supply chains.

These changes in supply chains have a direct impact on ports, as they modify goods flows and infrastructure requirements. Ports must adapt to these changes by improving their flexibility and capacity. For example, the increase in container transport by air, accelerated by the COVID-19 pandemic, requires investment in port infrastructures capable of handling growing freight volumes.

What's more, the rise of e-commerce and the increased demand for rapid delivery are prompting ports to rethink their operations. According to a McKinsey study, the global e-commerce market is expected to reach $6.4 trillion by 2024, creating opportunities for ports that can offer efficient logistics solutions.

In short, trends in international trade, including new trade routes and evolving supply chains, will shape the future of port infrastructures. To remain competitive, ports will need to adapt quickly to the changing global landscape, invest in modern infrastructure and develop collaborative strategies with other supply chain players. Integrating these trends into port planning will be key to meeting future challenges and maximizing growth opportunities.

Chapter 12: Sustainability and the environment

As the port sector evolves in the face of new economic and technological realities, environmental concerns are becoming inescapable. Ports are often major sources of pollution due to industrial activity, ship emissions and logistics infrastructures. This chapter explores strategies for ports to adopt environmentally-friendly practices and reduce their carbon footprint, thereby contributing to environmental sustainability.

1. Integrating environmental concerns

With increasing international regulations on greenhouse gas (GHG) emissions and the preservation of marine resources, ports need to integrate green strategies into their operations. For example, the International Maritime Organization (IMO) has set the target of reducing GHG emissions from the maritime sector by 50% by 2050. This pressure is driving ports to adopt environmentally-friendly practices to meet standards while remaining competitive.

Some ports, such as Rotterdam, have already implemented ambitious initiatives to reduce their environmental impact. This Dutch port has invested in renewable energies, notably wind and solar power, and encourages ships to use less polluting fuel by offering financial incentives. In addition, the development of "green ports", which use clean technologies and strive to reduce their environmental impact, is gaining ground in Europe and Asia. This approach includes the installation of electric terminals for ships at berth, thus reducing diesel-related emissions.

2. Reducing carbon footprint

Ports must also adopt specific practices to reduce their carbon footprint and minimize water and air pollution. One key strategy is to improve the energy efficiency of port equipment. For example, converting cranes and other handling equipment to electric operation helps reduce emissions of CO_2 and other atmospheric pollutants.

The electrification of port operations is a rapidly expanding field. For example, the Port of Los Angeles has implemented an electrification policy for its infrastructure, contributing to a 32% reduction in GHG emissions in ten years. At the same time, the use of carbon capture and storage (CCS) technologies is another solution being considered to reduce the port's carbon footprint.

Finally, the implementation of waste and ballast water management practices is an effective measure for preserving local ecosystems. The ports of Singapore and Hamburg are particularly advanced in this field, with integrated waste management and ballast water treatment systems to prevent contamination of aquatic environments.

In short, integrating sustainable practices and reducing carbon footprints are no longer just options, but imperatives for ports wishing to remain competitive and environmentally responsible. These efforts contribute to the fight against climate change and encourage more rational use of natural resources, promoting an environmentally-friendly model of port development.

Chapter 13: Technological developments

Technological advances are redefining port operations by introducing digital solutions and automation tools, which increase efficiency, reduce costs and improve safety. In this chapter, we look at how digitization and automation, including the use of drones, are transforming port management and optimizing logistics flows.

1. Digitization of port operations

Digitization represents one of the most significant transformations in modern ports, with technologies such as the Internet of Things (IoT), artificial intelligence (AI) and big data enabling real-time management of flows and resources. Digitization makes it possible to centralize and automate information on port traffic, storage and cargo handling, contributing to better decision-making and reduced waiting times for ships.

For example, the Port of Hamburg has integrated an IoT-based digital platform to monitor traffic and infrastructure in real time. This platform optimizes container storage and reduces waiting times by coordinating incoming and outgoing flows. In 2020, this approach reduced average ship handling time by 15%, demonstrating the positive impact of digitization on operational efficiency.

Digitization also contributes to more sustainable resource management. The Port of Rotterdam uses predictive analysis models to optimize infrastructure utilization, thereby reducing energy costs and cutting CO_2 emissions.

2. Automation and drones

The automation of port infrastructures, notably through the use of robots and autonomous vehicles, considerably improves the precision and speed of operations. Automation applies not only to container handling, but also to control and security processes. Ports such as Singapore have introduced robots to manage container loading and unloading, reducing dependence on human labor and increasing throughput.

Drones, meanwhile, are playing a growing role in the surveillance and inspection of port facilities. Used to inspect infrastructure, monitor flows and even assist security operations, drones save money and increase safety. At the Port of Los Angeles, drones are used to monitor cargo movements in real time and detect any suspicious activity. This reduces security costs and improves team responsiveness.

Finally, the use of autonomous vehicles for internal goods transport, such as the autonomous trucks at the port of Shanghai, demonstrates the potential of these technologies to reduce logistics times and costs. This automation also helps to minimize the risk of accidents and maintain a high work rate, even during peak periods.

All in all, technological developments in ports offer unprecedented prospects for optimization, making operations more efficient and meeting growing needs for safety and sustainability. The adoption of these technologies does, however, require significant investment and adaptation of infrastructures, but the benefits for the port sector as a whole justify these efforts.

General conclusion

In this conclusion, we summarize the crucial issues involved in managing port overcapacity and underactivity, highlighting the importance of balanced, proactive port infrastructure management. This final chapter also aims to provide concrete recommendations to guide decision-makers and investors towards sustainable solutions.

1. Summary of issues

Port overcapacity and underactivity pose major economic, logistical and environmental challenges, illustrating the complexity of the necessary balance between infrastructure supply and actual demand for port services. Overcapacity generates high maintenance costs and reduces the profitability of under-utilized facilities, while underactivity negatively impacts port competitiveness and dampens local economies. Both aspects show that hasty investment decisions, often motivated by regional competition or ill-adjusted anticipation of demand, can have significant long-term repercussions.

To meet these challenges, strategic management based on accurate forecasting, collaborative partnerships and adaptation to new business trends is essential. Initiatives such as digitization, automation and the implementation of sustainable practices can play a central role in optimizing the use of port infrastructures.

2. Call to Action

Faced with these challenges, it is crucial that governments, port managers and investors adopt a long-term vision, based on hard data and in-depth analyses of global needs and trends. Key recommendations include:

- **Adapted and flexible planning**: Planning in line with trade trends and regional needs can avoid errors of anticipation and limit the construction of unprofitable infrastructure.

- **Promoting public-private partnerships**: Collaborations between public and private players can encourage optimal use of resources and share the risks associated with investments.

- **Integration of technological innovations**: The adoption of advanced technologies, such as digital management systems and automation, can reduce operating costs, increase flow efficiency, and improve port competitiveness.

Finally, it is recommended to develop public policies encouraging sustainability, through fiscal incentives and regulations targeting responsible management of port resources. By applying these strategies, decision-makers can better anticipate fluctuating needs and strengthen the resilience of port infrastructures in the face of future challenges.

Contents

**Introduction

- Introduction to the subject**: This chapter introduces the context of the growth in port infrastructure worldwide, and the crucial role played by ports in the global and regional economy, as nerve centers for international trade.

- **Objectives of the book**: The main objectives are to unpack the challenges of port overcapacity and underactivity, and to propose sustainable solutions for efficient infrastructure management.

- **Methodology**: The analytical approach integrates case studies and global data to support proposed trends and analyses.

Part I: Background and importance of ports

**Chapter 1: Strategic role in international trade

- Explains the key role played by ports in the global logistics chain, as distribution and transshipment centers that are crucial to the smooth flow of international trade.

Chapter 2: Economic impact on the surrounding regions (in French)

- Analyzes the economic impact of ports on local employment and investment, as well as the direct and indirect impact of their activities on regional economies.

Part II: Definition of Concepts

**Chapter 3: Overcapacity

- Defines overcapacity as an excess of infrastructure over actual demand, detailing the types of overcapacity, including unprofitable facilities.

Chapter 4: Sub-activity

- Presents under-activity as a reduced use of port infrastructures, exploring its causes and consequences on profitability.

Part III: Issues

Chapter 5: Central question

- The main question is how overcapacity and underactivity affect the operational performance and profitability of ports, laying the foundations for the book's reflections and analyses.

Part I: Analysis of Port Overcapacity

Chapter 6: Causes of overcapacity (in French)

- Examines the factors leading to overcapacity, such as massive investments, the race for competitiveness between ports, and forecasts of commercial demand that don't always materialize.

Chapter 7: Consequences of overcapacity (in French)

- Identifies the high maintenance costs, declining profitability and environmental impacts of overcapacity, with illustrative examples.

**Chapter 8: Case studies

- Examines concrete examples of ports with overcapacity in Europe, America, Africa and Asia, comparing them with ports that have managed their capacity optimally.

Part II: Analysis of Port Sub-Activity

Chapter 9: Causes of Under-activity (in French)

- Analyzes how changing trade routes, competition and logistical obstacles can lead to under-utilization of port infrastructures.

Chapter 10: Consequences of Sub-Activity (in French)

- Explains how under-activity affects port competitiveness, results in local job losses and can lead to closures or redeployment.

**Chapter 11: Case studies

- Presents case studies of ports affected by under-activity, describing the measures taken to revitalize these infrastructures.

Part III: Solutions and Improvement Strategies

**Chapter 12: Infrastructure Optimization

- Discuss ways of reconfiguring facilities and making targeted investments to adapt capacity to real needs and reduce unnecessary costs.

**Chapter 13: Collaboration and partnerships

- Explore synergies between regional ports and public-private partnerships to optimize resources and develop joint port projects.

**Chapter 14: Technological Innovation

- Analyzes the benefits of digitalization and automation to improve efficiency, and explains how using data helps anticipate fluctuations in demand.

Chapter 15: Public policies and regulations (in French)

- Highlights the role of governments in regulating the port sector and tax incentives to attract investment in under-utilized ports.

Part IV: Future prospects

**Chapter 16: International trade trends

- Examines new trade routes, such as the Belt and Road, and their potential impact on ports, taking into account geopolitical developments and global supply chains.

**Chapter 17: Sustainability and the environment

- Explores the importance of green practices for ports, including reducing carbon footprints and adopting sustainable strategies in port operations.

Chapter 18: Technological evolutions (in French)

- Presents innovations such as digitization and the use of drones, offering solutions for more efficient and automated management of port operations.

Conclusion

**Summary of the Issues

- Summarizes the challenges of overcapacity and underactivity, highlighting the importance of striking a balance to ensure efficient port infrastructures.

**Call to Action

- Offers concrete recommendations for policy-makers and investors, encouraging them to adopt port management practices that promote sustainable growth.

Bibliography and Appendices

- **Bibliography**: Comprehensive list of books, academic articles, reports from international organizations, and online sources used to support the analyses in this book.

- Appendices**: Includes graphs, maps and tables of statistical data to provide detailed visual support for case studies and analyses, and to help you delve deeper into certain technical aspects of the book.

This summary provides a structured overview of the key themes and in-depth analyses that will be covered, making it easy to navigate through port infrastructure issues and solutions.

Sources, References and Bibliography

This book draws on a variety of sources and references, offering a comprehensive overview of global port issues. The following elements illustrate the wealth of documentation mobilized to support the analyses and proposals.

1.**Scientific Articles and Academic Publications**)
-Port management and logistics studies: trade journals such as *Journal of Transport Geography*, *Maritime Policy & Management*, and *International Journal of Shipping and Transport Logistics*.
-Academic articles on the economic impact of ports on surrounding regions.

2 Reports and Official Publications** (in French)
-International Maritime Organization (IMO) reports on sustainable port development.
World Bank and Organisation for Economic Co-operation and Development (OECD) documents infrastructure and international trade.
-International Transport Forum reports on shipping route trends and port capacity.

3.**Specialized works
-**Rodrigue, J.-P. (2020). *The Geography of Transport Systems*.**
This book provides a solid theoretical basis for logistics and transport infrastructures.

- Notteboom, T., & Winkelmans, W. (2009). *A key book on port management strategies.

- **Brooks, M. R., & Cullinane, K. (2017). *Devolution, Port Governance, and Port Performance*.** An analysis of port governance patterns and their impact on performance.

4. **Case studies

- Concrete examples drawn from analysis of ports such as Rotterdam, Singapore, Shanghai and emerging African ports such as Kribi (Cameroon).

- Port revitalization initiatives in Asia and Latin America.

5. **Statistical sources and data

- United Nations Conference on Trade and Development (UNCTAD) database for maritime traffic statistics.

- Information from the European Environment Agency on the environmental impact of port infrastructures.

6. **Press and media articles

- Financial Times* and *The Economist* articles on trade routes and port investments.

- Analyses from shipping media such as *Lloyd's List* and *Port Technology International*.

7. **Government Reports and National Publications**.

- National and regional port policies, available in the reports of the Ministries of Transport of various countries.

- Case studies from the National Port Infrastructure Agency's publications in countries such as Cameroon and Kenya.

This bibliography and set of sources help to structure a robust analytical framework, supporting each chapter with empirical evidence and recognized analyses.

Printed by Books on Demand GmbH, Norderstedt / Germany